Beyond the Standard Model: High-Energy Particle Physics for Every Curious Mind

Petrova

Copyright © [2023]

Title: Beyond the Standard Model: High-Energy Particle Physics for Every Curious Mind

Author's: Petrova.

This book was printed and published by [Publisher's: Petrova]   in [2023]

ISBN:

TABLE OF CONTENTS

Recapitulation of Key Concepts

The Impact of High-Energy Particle Physics

Chapter 1: Introduction to High-Energy Particle Physics

The Standard Model: A Brief Overview

Welcome to the exciting world of particle physics! In this subchapter, we will delve into the fascinating topic of the Standard Model—a fundamental theory that has revolutionized our understanding of the universe. Whether you are an enthusiast, a student, or simply curious about the wonders of the cosmos, this brief overview will provide you with a solid foundation to appreciate the intricacies of particle physics.

The Standard Model is a remarkable framework that describes the fundamental building blocks of matter and their interactions. It encompasses three of the four fundamental forces of nature: electromagnetism, the weak nuclear force, and the strong nuclear force. This comprehensive theory has been meticulously developed over the course of several decades, combining the efforts of brilliant minds and countless experimental observations.

At its core, the Standard Model consists of two main classes of particles: fermions and bosons. Fermions are the building blocks of matter and include particles such as quarks and leptons. Quarks are the constituents of protons and neutrons, while leptons include familiar particles like electrons and neutrinos. These fermions interact through the exchange of bosons, which are force-carrying particles. For instance, photons mediate electromagnetic interactions, while W and Z bosons are responsible for the weak nuclear force.

One of the most intriguing aspects of the Standard Model is its prediction of the Higgs boson—a particle that gained substantial attention with the discovery at CERN's Large Hadron Collider in 2012. The Higgs boson is associated with the Higgs field, which permeates the universe and endows particles with mass. Its discovery was a significant milestone, confirming a key component of the Standard Model and shedding light on the origin of mass in the universe.

While the Standard Model has been incredibly successful in explaining a vast array of experimental observations, it is not without its limitations. For example, it does not incorporate gravity, which is described by Einstein's general theory of relativity. Additionally, the existence of dark matter and dark energy remains unexplained within the framework of the Standard Model.

In conclusion, the Standard Model is a remarkable achievement of human intellect, providing us with a comprehensive understanding of the fundamental particles and forces that shape our universe. Although it has some shortcomings, it has stood the test of time and serves as an essential foundation for exploring the mysteries of particle physics. So, join us on this incredible journey, as we venture beyond the Standard Model and delve into the frontiers of high-energy particle physics.

Limitations of the Standard Model

The Standard Model of particle physics has been an incredibly successful theory in describing the fundamental particles and their interactions. However, like any scientific theory, it has its limitations. In this subchapter, we explore some of the key shortcomings of the Standard Model, shedding light on the mysteries that continue to captivate the field of particle physics.

One major limitation of the Standard Model is its inability to account for gravity. Despite successfully explaining the electromagnetic, weak, and strong forces, the theory falls short when it comes to incorporating gravity into its framework. This has been a long-standing challenge in the field, with physicists striving to develop a unified theory that encompasses all fundamental forces, including gravity.

Another limitation lies in the inability of the Standard Model to explain dark matter, which is believed to make up a significant portion of the universe's mass. While the theory provides an excellent description of visible matter, it fails to account for the existence of dark matter, which only interacts weakly with ordinary matter. Unraveling the nature of dark matter is one of the most pressing questions in modern particle physics.

Furthermore, the Standard Model cannot explain the matter-antimatter asymmetry observed in the universe. According to the theory, matter and antimatter should have been created in equal amounts during the early stages of the universe. However, our observations show a significant excess of matter over antimatter.

Understanding this asymmetry, known as the baryon asymmetry problem, remains a key challenge for physicists.

Additionally, the Standard Model lacks a mechanism to explain the masses of neutrinos. While the theory successfully predicts the masses of other elementary particles, neutrinos have been observed to have extremely small but non-zero masses. Exploring the origin of neutrino masses and their implications for particle physics is an active area of research.

In conclusion, while the Standard Model has been an exceptional theory in explaining the fundamental particles and their interactions, it faces several limitations. The absence of a unifying theory that incorporates gravity, the mystery of dark matter, the matter-antimatter asymmetry, and the small but non-zero masses of neutrinos are some of the key challenges that particle physicists continue to grapple with. By acknowledging and exploring these limitations, we pave the way for new breakthroughs and a deeper understanding of the fundamental nature of our universe.

Importance of High-Energy Particle Physics

Particle physics, also known as high-energy physics, is a fascinating field that seeks to understand the fundamental building blocks of the universe and the forces that govern them. It delves into the deepest layers of reality, exploring the nature of matter, energy, space, and time. The importance of high-energy particle physics cannot be overstated, as it plays a crucial role in shaping our understanding of the universe and has numerous practical applications.

One of the key contributions of high-energy particle physics is its ability to provide insights into the fundamental laws of nature. By studying the behavior of subatomic particles at extremely high energies, scientists can uncover new phenomena and test existing theories, such as the widely accepted Standard Model. This model describes the known particles and their interactions but is not a complete picture of the universe. High-energy particle physics aims to go beyond the Standard Model, searching for new particles, forces, and dimensions that could revolutionize our understanding of the cosmos.

Another reason why high-energy particle physics is important lies in its technological and societal impact. The cutting-edge technologies developed for particle accelerators and detectors have led to numerous advancements in various industries. For instance, medical imaging techniques like PET scans and radiation therapy have benefited from the innovations in particle physics. Additionally, the data analysis and computational techniques developed in this field have found applications in fields as diverse as finance, weather forecasting, and data security.

Furthermore, high-energy particle physics fosters international collaboration and cooperation. Large-scale experiments, such as the Large Hadron Collider (LHC), bring together scientists from different countries, cultures, and backgrounds, working towards a common goal. This collaboration not only accelerates scientific progress but also promotes cultural exchange and understanding.

Moreover, high-energy particle physics has the potential to address some of the most pressing questions in science and society. It can shed light on the mysteries of dark matter and dark energy, which together constitute about 95% of the universe. Understanding these enigmatic entities could have far-reaching implications for cosmology, astrophysics, and our understanding of the origins of the universe.

In conclusion, high-energy particle physics is of paramount importance to our understanding of the universe, as it allows us to explore the fundamental laws of nature, develop new technologies, foster international collaboration, and address profound scientific questions. Whether you are a curious individual exploring the wonders of the universe or a particle physics enthusiast, delving into the realm of high-energy particle physics will undoubtedly expand your knowledge and curiosity about the world we inhabit.

Historical Milestones in High-Energy Particle Physics

In the vast realm of high-energy particle physics, numerous groundbreaking discoveries have shaped our understanding of the fundamental building blocks of the universe. These historical milestones have paved the way for remarkable advancements in our knowledge and have led to the development of the Standard Model, the current framework that describes the elementary particles and their interactions. Let us embark on a journey through time, exploring the key moments that have shaped the field of particle physics.

One of the earliest milestones in this field was the discovery of the electron by J.J. Thomson in 1897. This tiny, negatively charged particle was the first indication that matter is composed of smaller constituents. This discovery laid the foundation for further investigations into particle physics.

In the early 20th century, the work of Ernest Rutherford and his colleagues revolutionized our understanding of atomic structure. Rutherford's famous gold foil experiment led to the discovery of the atomic nucleus and the realization that most of an atom's mass is concentrated in a tiny, positively charged region. This discovery opened up new avenues for the study of particle interactions and the forces that govern them.

The 1930s marked the discovery of the positron, the antimatter counterpart of the electron, by Carl Anderson. This revelation shattered the notion of matter being solely composed of particles with positive mass. It laid the groundwork for the development of quantum electrodynamics (QED), the first quantum field theory that successfully described the behavior of particles and their interactions via the electromagnetic force.

Another significant milestone occurred in 1964 when Murray Gell-Mann and George Zweig independently proposed the concept of quarks. Quarks are the fundamental constituents of protons, neutrons, and other hadrons. This discovery introduced the concept of "color charge" and led to the development of quantum chromodynamics (QCD), the theory that describes the strong nuclear force.

Perhaps the most iconic milestone in high-energy particle physics was the discovery of the Higgs boson in 2012 at CERN's Large Hadron Collider (LHC). This elusive particle, postulated by Peter Higgs and others, confirmed the existence of the Higgs field and its role in endowing other particles with mass. The discovery of the Higgs boson completed the Standard Model and validated decades of theoretical and experimental efforts.

These historical milestones highlight the remarkable progress made in high-energy particle physics, revealing the intricate nature of the universe at its most fundamental level. They serve as a testament to the curiosity and determination of scientists who continue to push the boundaries of our knowledge. As we delve deeper into the mysteries of the cosmos, these milestones will undoubtedly be stepping stones towards a new era of discovery beyond the Standard Model.

Chapter 2: Fundamental Particles and Interactions

Elementary Particles: Building Blocks of Matter

In the vast expanse of the universe, everything we see, touch, and interact with is made up of tiny, fundamental entities called elementary particles. These minuscule building blocks of matter are the focus of study in the fascinating field of particle physics. In this subchapter, we will delve into the intricate world of elementary particles and explore how they shape the very fabric of our universe.

To understand the nature of elementary particles, we must first comprehend the fundamental forces that govern their behavior. These forces, namely gravity, electromagnetism, the weak nuclear force, and the strong nuclear force, act as the guiding principles behind the interactions of particles. Each force is mediated by specific particles, known as force carriers or gauge bosons, which enable the transfer of energy between elementary particles.

The Standard Model of particle physics provides a framework for organizing and categorizing these elementary particles. It classifies them into two main groups: fermions and bosons. Fermions, such as quarks and leptons, form the basis of matter and are responsible for its diverse properties. On the other hand, bosons, including the force carriers, mediate the interactions between particles, allowing them to come together or repel each other.

Quarks, which come in six different flavors, combine to form protons and neutrons, the building blocks of atomic nuclei. Leptons, such as electrons and neutrinos, are elementary particles that exist independently and are not affected by the strong nuclear force. These particles interact through the electromagnetic and weak nuclear forces, playing crucial roles in the stability and behavior of matter.

Additionally, the Higgs boson, discovered in 2012 at the Large Hadron Collider, is a vital piece of the puzzle. It is responsible for giving mass to elementary particles, allowing them to acquire the properties we observe in the physical world.

Understanding elementary particles is not only a matter of scientific curiosity; it has practical implications as well. Advances in particle physics research have led to breakthrough technologies in medical imaging, energy production, and telecommunications, among others. By unraveling the mysteries of these tiny entities, scientists hope to unlock new frontiers of knowledge and revolutionize our understanding of the universe.

Whether you are a curious mind seeking to expand your knowledge or a physics enthusiast eager to explore the depths of particle physics, the study of elementary particles opens a gateway to a captivating realm where the tiniest entities shape the grandest phenomena. Join us as we embark on a journey beyond the Standard Model, unraveling the secrets of high-energy particle physics that lie at the heart of our existence.

Fermions: Quarks and Leptons

In the fascinating world of particle physics, the study of fermions—quarks and leptons—holds the key to understanding the fundamental building blocks of our universe. In this subchapter, we delve into the intricacies of these subatomic particles, shedding light on their properties, interactions, and the mysteries they unlock.

Quarks, the ever-elusive constituents of protons and neutrons, are the true essence of matter. They come in six different flavors: up, down, charm, strange, top, and bottom. Each quark possesses a fractional electric charge, making them the only particles to exhibit such an extraordinary feature. Quarks are never found in isolation due to their confinement within composite particles called hadrons. The strong nuclear force, mediated by gluons, binds quarks together, creating a captivating dance of color charges. Understanding the behavior of quarks has been a monumental task for physicists, resulting in the development of quantum chromodynamics (QCD), a theory that describes the strong interaction.

On the other hand, leptons are the elementary particles that do not experience the strong nuclear force. Electrons, muons, and taus, along with their corresponding neutrinos, comprise the lepton family. Leptons possess an electric charge and interact via the electromagnetic and weak nuclear forces. Neutrinos, however, are unique among leptons as they are electrically neutral and only weakly interact with matter. Their enigmatic nature has captivated scientists for decades, leading to groundbreaking discoveries such as neutrino oscillations, which have revolutionized our understanding of particle physics.

Beyond their individual properties, the significance of fermions lies in their role as matter carriers. Quarks and leptons combine to form the atoms and molecules that constitute our everyday world. The Standard Model, the reigning theory of particle physics, provides a framework for understanding the behavior of these particles and their interactions. However, it falls short in explaining some profound phenomena, such as dark matter, dark energy, and the matter-antimatter asymmetry in the universe.

As we venture beyond the Standard Model, exploring the realms of high-energy particle physics, we are propelled into a world of unanswered questions and tantalizing possibilities. From the mysteries of mass generation to the quest for unification, researchers are tirelessly searching for clues that lie beyond the reach of our current understanding.

Whether you are an avid enthusiast or simply curious about the secrets of the universe, this subchapter on Fermions: Quarks and Leptons will unravel the complex world of particle physics. Join us on this exhilarating journey as we delve into the depths of matter and energy, uncovering the hidden truths that shape our cosmic existence.

Gauge Bosons: Carriers of Fundamental Forces

In the fascinating world of particle physics, scientists have discovered that the universe is governed by four fundamental forces: gravity, electromagnetism, the weak nuclear force, and the strong nuclear force. These forces are responsible for all interactions between particles and play a crucial role in shaping the universe as we know it. But have you ever wondered how these forces are transmitted across vast distances? Enter gauge bosons, the unsung heroes of the particle world.

Gauge bosons are the carriers of these fundamental forces, acting as the mediators between particles. Just as messengers deliver information from one person to another, gauge bosons transmit the forces between particles, allowing them to interact and exchange energy.

Let's start with the electromagnetic force, the force responsible for everything from the light we see to the electrical signals in our brains. The gauge boson associated with this force is the photon, a massless particle that travels at the speed of light. Photons enable the interaction between charged particles, such as electrons, by carrying the electromagnetic force between them.

Moving on to the weak nuclear force, responsible for processes like radioactive decay and the fusion of hydrogen in the Sun. This force is mediated by three gauge bosons: W+, W-, and Z. Unlike the photon, these gauge bosons have mass, which gives them a shorter range of interaction. Nevertheless, they play a crucial role in enabling interactions between particles, allowing the weak force to shape the behavior of matter on a subatomic level.

Next, we have the strong nuclear force, which holds atomic nuclei together. Gluons, the gauge bosons associated with this force, are unique in that they carry the force itself. Unlike other gauge bosons, gluons interact with each other, resulting in the confinement of quarks within protons and neutrons. This force is responsible for the stability of matter as we know it.

Lastly, we come to gravity, the force that governs the motion of celestial bodies and shapes the structure of the universe. While the other forces have their gauge bosons, gravity is a bit different. In the Standard Model of particle physics, the theoretical framework that describes the known particles and forces, the gauge boson for gravity, known as the graviton, has yet to be discovered. However, scientists continue to search for evidence of its existence, hoping to unite gravity with the other fundamental forces.

Understanding gauge bosons and their role as carriers of fundamental forces is essential in unraveling the mysteries of the universe. As we delve deeper into particle physics, we gain a greater appreciation for the intricate dance of these invisible messengers, shaping the very fabric of our existence.

Higgs Boson: The Particle of Mass

In the fascinating world of particle physics, there exists a tiny yet groundbreaking particle known as the Higgs boson. Often referred to as the "God particle," the Higgs boson plays a crucial role in our understanding of mass and the fundamental forces that govern our universe. This subchapter delves into the captivating realm of the Higgs boson and its significance in the field of particle physics.

First postulated by physicist Peter Higgs in the 1960s, the Higgs boson remained a theoretical entity until its discovery at the Large Hadron Collider (LHC) in 2012. The quest for this elusive particle involved decades of research and the collaboration of thousands of scientists from around the world. Its discovery was a monumental achievement, confirming the existence of the Higgs field, which permeates all of space and interacts with particles to grant them mass.

To understand how the Higgs boson imparts mass, we must first grasp the concept of the Higgs field. Much like an invisible cosmic molasses, the Higgs field fills the vacuum of space. When particles move through this field, they experience resistance, akin to wading through a pool of water. This resistance slows down some particles, endowing them with mass, while others glide through unaffected, remaining massless.

The Higgs boson emerges as a byproduct of the Higgs field. Its discovery was a crucial piece of the puzzle, confirming the existence of the field and validating the Standard Model of particle physics. This model describes the fundamental particles and forces of nature, and the Higgs boson completes the picture by explaining why particles have mass.

Furthermore, the discovery of the Higgs boson has opened up new avenues for research, pushing the boundaries of our understanding of the universe. Scientists are now exploring the properties of the Higgs boson, such as its decay modes and interactions with other particles. These investigations may provide insights into the nature of dark matter, the origin of the universe, and even the possibility of additional dimensions.

The Higgs boson truly stands as a remarkable particle, revolutionizing our understanding of mass and the fundamental fabric of our universe. Its discovery represents a triumph of human curiosity and scientific collaboration. As we continue to unravel the mysteries of particle physics, the Higgs boson remains a beacon of knowledge, guiding us towards a deeper understanding of the cosmos and our place within it.

Understanding Interactions: Quantum Field Theory

In the fascinating world of particle physics, the study of interactions between fundamental particles is crucial to unraveling the mysteries of the universe. One powerful framework that helps us comprehend these interactions is Quantum Field Theory (QFT). This subchapter aims to introduce the concept of QFT to everyone interested in particle physics, providing a glimpse into the remarkable world of interactions at the subatomic level.

Quantum Field Theory revolutionized our understanding of particles and their interactions, merging quantum mechanics with special relativity. It describes particles not as individual entities but as excitations of underlying fields that permeate all of space and time. These fields are quantized, meaning they can only exist in discrete energy states, giving rise to particles that we observe in experiments.

The fundamental principle of QFT is the concept of a field, which is a mathematical entity that assigns values to each point in space and time. These fields can be classified into different types, such as the electromagnetic field or the Higgs field. Each field has its corresponding particle, such as the photon for the electromagnetic field or the Higgs boson for the Higgs field.

Interactions between particles, mediated by these fields, are the building blocks of our universe. They provide the means for particles to exchange energy, momentum, and other fundamental properties. Understanding these interactions is essential to deciphering the nature of matter and the fundamental forces that govern the universe.

QFT also introduces the concept of particles and antiparticles, which are necessary for a complete description of interactions. Antiparticles are essentially mirror images of particles, possessing the same mass but opposite charge. The existence of antiparticles is a consequence of the symmetries inherent in QFT.

By treating particles and fields as quantum objects, Quantum Field Theory allows us to calculate the probabilities of various interactions occurring. These calculations involve complex mathematical techniques, such as Feynman diagrams, which provide a graphical representation of particle interactions.

The study of Quantum Field Theory has led to numerous breakthroughs in particle physics, including the development of the Standard Model, our current best description of the fundamental particles and their interactions. However, the Standard Model is not a complete theory and leaves many questions unanswered, such as the nature of dark matter or the unification of all fundamental forces.

In conclusion, Quantum Field Theory is a powerful framework that enables us to understand the interactions between particles, shedding light on the fundamental nature of our universe. By grasping the concepts of fields, particles, and their interactions, we embark on an exciting journey into the realm of particle physics, where we strive to unravel the mysteries that lie beyond the Standard Model.

Chapter 3: Particle Accelerators and Detectors

Introduction to Particle Accelerators

Particle accelerators are fascinating machines that have revolutionized the field of particle physics. They are at the forefront of scientific research, allowing us to delve deep into the mysteries of the subatomic world. In this subchapter, we will explore the fundamental concepts and workings of particle accelerators, shedding light on their importance in advancing our understanding of the universe.

At its core, a particle accelerator is a powerful tool used to accelerate charged particles, such as protons or electrons, to incredibly high speeds. By doing so, scientists can recreate extreme conditions that mimic the early moments of the universe or probe the fundamental building blocks of matter. These accelerated particles collide with each other or stationary targets, generating new particles and enabling the study of their properties.

The design and construction of particle accelerators require a combination of engineering, physics, and cutting-edge technology. They come in various shapes and sizes, ranging from small tabletop accelerators used in medical diagnostics to massive, multi-kilometer-long colliders like the Large Hadron Collider (LHC) at CERN. The LHC, for instance, is the largest and most powerful accelerator ever built, allowing scientists to explore energies never before reached in a laboratory.

Accelerators can be classified into two main types: linear accelerators (linacs) and circular accelerators (synchrotrons). Linacs accelerate particles in a straight line using electric fields, while synchrotrons use magnetic fields to bend the particles in a circular path, continuously increasing their kinetic energy. Both types have their advantages and are used for different scientific purposes.

One of the key applications of particle accelerators is the discovery and study of new particles. For example, the Higgs boson, a particle that plays a crucial role in the mechanism of mass, was discovered at the LHC in 2012. By colliding particles at high energies, scientists can explore uncharted territories, searching for new particles that could provide insights into the fundamental nature of the universe.

In addition to particle discovery, accelerators have practical applications in various fields. They are used in medicine for cancer treatment through radiotherapy, in materials science to study the behavior of materials under extreme conditions, and in industry for materials testing and quality control.

Particle accelerators truly represent the cutting edge of scientific research, pushing the boundaries of our knowledge and understanding of the universe. They allow us to unravel the secrets of the smallest particles and explore the fundamental forces that govern our world. In the following chapters, we will delve deeper into the workings of particle accelerators, their components, and their role in unraveling the mysteries of particle physics. So, join us on this exciting journey as we explore the fascinating world of particle accelerators and the mysteries they help us unlock.

Linear Accelerators

In the fascinating world of particle physics, linear accelerators play a crucial role in investigating the fundamental building blocks of the universe. These incredible machines propel charged particles to incredibly high speeds, enabling scientists to study the intricacies of matter and the forces that govern our universe. In this subchapter, we will delve into the inner workings of linear accelerators and their significance in pushing the boundaries of our understanding.

Linear accelerators, often referred to as linacs, are devices that accelerate charged particles along a straight path. Unlike circular accelerators such as the Large Hadron Collider, linacs operate in a linear fashion, propelling particles forward in a highly controlled manner. This design allows for precise control over the particle's path and energy, making linacs indispensable tools for particle physicists.

The basic principle behind linear accelerators is the application of electric fields to accelerate and propel particles forward. By alternating the polarity of these electric fields, particles gain energy as they pass through each segment of the accelerator. As the particles progress through the linac, their speed increases until they reach their desired energy level.

One of the primary advantages of linear accelerators is their versatility. They can accelerate a wide range of particles, from electrons and protons to heavier ions, providing invaluable insights into different aspects of particle physics. Additionally, linacs can produce intense and focused particle beams, allowing for precise and targeted experiments.

Linear accelerators have been instrumental in numerous breakthroughs in particle physics. They have been used to study the properties and interactions of subatomic particles, shedding light on the fundamental forces that shape our universe. Linacs have also played a vital role in medical applications, such as cancer treatment through radiation therapy.

The development and improvement of linear accelerators continue to be areas of active research. Scientists are constantly striving to enhance their performance, increase particle energies, and explore new frontiers of particle physics. As technology progresses, linear accelerators are becoming more compact, efficient, and affordable, making them accessible to a wider range of researchers and institutions.

In conclusion, linear accelerators are remarkable machines that have revolutionized our understanding of the universe. They have enabled scientists to delve deeper into the mysteries of particle physics, uncovering the secrets of matter, energy, and the forces that govern our existence. With their versatility and precision, linear accelerators continue to push the boundaries of scientific knowledge, opening up new avenues for exploration and discovery. Whether you are a curious mind or a particle physics enthusiast, understanding the significance of linear accelerators is essential to appreciating the remarkable advancements in this field.

Circular Accelerators

Circular accelerators play a crucial role in the field of particle physics, allowing scientists to study the fundamental building blocks of the universe in great detail. These powerful machines are designed to accelerate subatomic particles to extremely high speeds, enabling them to collide and create new particles, revealing the secrets of the universe at the smallest scales. In this subchapter, we will delve into the fascinating world of circular accelerators and explore how they have revolutionized our understanding of particle physics.

Circular accelerators, as the name suggests, consist of a circular ring-shaped path along which particles are accelerated. These rings are equipped with powerful magnets that bend the path of the particles, keeping them confined to the circular trajectory. The particles gain energy with each lap around the ring, thanks to electromagnetic fields that accelerate them to higher speeds. This process continues until the particles reach the desired energy levels for experiments.

One of the most well-known examples of a circular accelerator is the Large Hadron Collider (LHC) at CERN. The LHC is a 27-kilometer ring located underground on the border of France and Switzerland. It accelerates protons to nearly the speed of light and then collides them head-on, creating conditions similar to those just after the Big Bang. These collisions generate an enormous amount of energy, which can be used to produce heavy and short-lived particles that would not exist under normal conditions.

Circular accelerators have led to countless discoveries in particle physics. By colliding particles at high speeds, scientists can study the resulting debris and identify new particles that were previously unknown. These discoveries have expanded our understanding of the universe and its fundamental forces.

Moreover, circular accelerators have practical applications beyond fundamental research. They are used in medical facilities to produce beams of high-energy particles for cancer treatment, delivering precise doses of radiation to tumors while minimizing damage to healthy tissue. This technology has revolutionized cancer treatment and given hope to millions of patients worldwide.

In conclusion, circular accelerators are remarkable machines that have revolutionized our understanding of particle physics. They allow scientists to study the fundamental building blocks of the universe by colliding particles at high speeds. These accelerators have not only expanded our knowledge of the universe but also found practical applications in fields like medicine. By harnessing the power of circular accelerators, scientists continue to push the boundaries of scientific knowledge and unlock the secrets of the universe, inspiring awe and curiosity for everyone interested in particle physics.

Detectors: Capturing High-Energy Particle Interactions

In the fascinating world of high-energy particle physics, understanding the fundamental building blocks of matter and unraveling the mysteries of the universe requires sophisticated tools known as detectors. These remarkable devices play a crucial role in capturing and analyzing the interactions of high-energy particles, providing valuable insights into the fundamental laws that govern our universe.

When it comes to studying the intricate world of particle physics, detectors are the key to unlocking the hidden secrets of the subatomic realm. These detectors are designed to identify and measure various properties of particles, such as their energy, mass, charge, and momentum. By meticulously capturing and analyzing the interactions between particles, detectors allow scientists to reconstruct the intricate paths of these particles, providing valuable information about their properties and behavior.

One of the most commonly used detectors in particle physics is the particle collider. These gigantic machines, such as the Large Hadron Collider (LHC), accelerate particles to incredibly high speeds and collide them together. The collisions that occur within these detectors generate a plethora of new particles, often lasting for only a fraction of a second. By meticulously studying these collisions and the resulting particle showers, physicists can gain insights into the fundamental forces and particles that govern the universe.

Another type of detector commonly employed in high-energy particle physics is the calorimeter. Calorimeters are designed to measure the energy of particles that pass through them. These devices absorb the energy of the particles, creating a measurable signal that can be used to determine the type and energy of the particle. Calorimeters are particularly useful in measuring the energy of particles such as electrons, photons, and hadrons.

In addition to particle colliders and calorimeters, there are various other types of detectors used in high-energy particle physics, including trackers, time-of-flight detectors, and electromagnetic calorimeters. Each detector serves a specific purpose in capturing and analyzing different aspects of particle interactions, contributing to our understanding of the fundamental laws that govern the universe.

Detectors not only play a crucial role in advancing our knowledge of particle physics but also find applications in various other fields, such as medicine and industry. From cancer treatment to airport security scanners, the principles and technologies developed in high-energy particle physics have had far-reaching impacts on our daily lives.

In conclusion, detectors are vital tools in the field of high-energy particle physics. By capturing and analyzing the interactions of high-energy particles, these devices provide scientists with invaluable insights into the fundamental laws that govern our universe. With their wide range of applications and their ability to unlock the secrets of the subatomic realm, detectors have revolutionized our understanding of particle physics, making them an indispensable tool for every curious mind.

Experimental Techniques in High-Energy Physics

High-energy physics is a captivating field that delves into the fundamental building blocks of the universe, exploring the smallest scales of matter and energy. To uncover the mysteries of particle physics, scientists employ a variety of experimental techniques that push the boundaries of technology and our understanding of the universe. In this subchapter, we will explore some of the cutting-edge experimental techniques used in high-energy physics research.

One of the primary tools in high-energy physics research is particle accelerators. These colossal machines propel particles to incredibly high speeds, allowing scientists to study their behavior when they collide. Accelerators come in various forms, from linear accelerators that accelerate particles in a straight line to circular colliders that collide particles in a circular path. These collisions generate an enormous amount of energy, mimicking the conditions just after the Big Bang and providing insights into the fundamental forces and particles that govern our universe.

To detect and analyze the particles produced in these collisions, high-energy physicists employ sophisticated detectors. These detectors are designed to measure various properties of particles, such as their energy, charge, and momentum. Detectors often consist of layers of different materials, each serving a specific purpose. For example, the innermost layers may consist of silicon detectors that can track the paths of charged particles, while the outer layers may be calorimeters that measure the energy of particles by absorbing them.

Another important experimental technique is the use of particle collisions to create new particles. By colliding particles at high energies, scientists can produce particles that are not naturally present in our everyday environment. These newly created particles may be short-lived and decay into other particles, which can provide valuable insights into the underlying physics.

In recent years, high-energy physics experiments have also begun to exploit the power of large-scale collaborations. International collaborations, such as the ATLAS and CMS experiments at CERN's Large Hadron Collider, bring together scientists from all over the world. These collaborations allow for the pooling of resources, expertise, and data, enabling scientists to tackle complex questions that no single institution could address alone.

Experimental techniques in high-energy physics continue to evolve and improve, driven by the curiosity of scientists and the technological advancements of our time. These techniques not only deepen our understanding of the universe but also inspire breakthroughs in other fields, such as medicine and technology. As we venture into the unknown, the quest for knowledge in particle physics captivates the curious minds of scientists and enthusiasts alike, shaping our understanding of the fundamental nature of reality.

Chapter 4: Energy Scales and the Hierarchy Problem

Energy Scales in Fundamental Physics

In the ever-expanding field of particle physics, understanding energy scales is crucial to unraveling the mysteries of the universe. From the smallest subatomic particles to the immense energies found in the cosmos, energy scales play a vital role in providing insights into the fundamental workings of nature. This subchapter explores the fascinating world of energy scales in fundamental physics, taking you on a journey from the tiniest building blocks of matter to the grandest scales of the universe.

At the heart of particle physics lies the Standard Model, a theory that beautifully describes the interactions between elementary particles and three fundamental forces: electromagnetism, the weak force, and the strong force. However, the Standard Model is not without its limitations, and it fails to provide explanations for certain phenomena like dark matter and the hierarchy problem. To delve deeper into these mysteries, scientists have been working tirelessly to probe energy scales beyond those accessible by current experiments.

The energy scale at which particle physics experiments operate is typically measured in electron volts (eV). The familiar everyday energy scale, such as the energy of visible light, is around a few eV. However, particle physicists deal with much higher energy scales, spanning several orders of magnitude. For instance, the Large Hadron Collider (LHC), the world's most powerful particle accelerator, collides protons at energies around 14 trillion eV (TeV), allowing us to explore the subatomic world with unprecedented precision.

Beyond the energy scales accessible at the LHC, theorists speculate the existence of new particles and phenomena that could shed light on the mysteries of the universe. These hypothetical particles, often referred to as "new physics," might be interconnected with energy scales that are several orders of magnitude higher than those currently probed. By pushing the boundaries of our understanding, scientists hope to uncover the secrets hidden at these ultra-high energy scales.

Moreover, energy scales are not limited to the microscopic realm. In cosmology, the study of the universe on the largest scales, energies reach mind-boggling magnitudes. During the early moments of the universe, the energy scale was so high that particles and forces behaved differently than they do today. Understanding these primordial energy scales can provide valuable insights into the origin and evolution of our universe.

In conclusion, energy scales are central to the study of fundamental physics, both at the microscopic and cosmic levels. By exploring these scales, scientists aim to unravel the mysteries of the universe and pave the way for the discovery of new physics beyond the current understanding. Whether you're a curious enthusiast or a seasoned particle physicist, delving into the world of energy scales opens up a realm of excitement and wonder, challenging our understanding of the fundamental fabric of reality.

The Hierarchy Problem: Why is Gravity So Weak?

Gravity, the force that governs the motion of planets, stars, and galaxies, has always fascinated scientists and philosophers alike. However, when it comes to understanding the nature of gravity in the context of particle physics, a peculiar question arises: why is gravity so weak compared to the other fundamental forces?

In the subchapter titled "The Hierarchy Problem: Why is Gravity So Weak?" of the book "Beyond the Standard Model: High-Energy Particle Physics for Every Curious Mind," we delve into this intriguing phenomenon. This subchapter aims to provide a comprehensive explanation of the hierarchy problem, shedding light on why gravity appears so feeble in comparison to electromagnetism, the strong nuclear force, and the weak nuclear force.

To grasp the essence of the hierarchy problem, we must first understand the concept of energy scales. In the realm of particle physics, different interactions occur at specific energy scales. Gravity, being the weakest force, operates at an energy scale much larger than the other forces. This stark contrast in energy scales creates a conundrum as to why gravity is so feeble at the particle level.

The subchapter proceeds to explore various theoretical explanations for the hierarchy problem, including the role of extra dimensions, supersymmetry, and the Higgs mechanism. These concepts are introduced in an accessible manner, ensuring that readers from all backgrounds can comprehend the significance of these theories.

Moreover, the subchapter delves into the ramifications of the hierarchy problem. It discusses how this puzzle is intricately linked with the search for a theory of everything, which would unify all the fundamental forces. By understanding the hierarchy problem, scientists can potentially unlock the secrets of the universe and gain a deeper comprehension of the fundamental laws that govern our existence.

"The Hierarchy Problem: Why is Gravity So Weak?" is an essential subchapter for anyone intrigued by the fascinating field of particle physics. Whether you are a curious individual with a general interest in the subject or a devoted particle physicist, this section of the book offers an engaging exploration of the hierarchy problem. By unraveling the mysteries surrounding gravity's weakness, this subchapter takes readers on a thought-provoking journey, revealing the interconnectedness of the fundamental forces and the profound implications for our understanding of the universe.

Supersymmetry: A Solution to the Hierarchy Problem?

In the fascinating realm of particle physics, one of the most intriguing puzzles scientists have grappled with is known as the hierarchy problem. This conundrum revolves around the vast difference in energy scales between gravity and the other fundamental forces of nature. Why is gravity so much weaker than the other forces? Is there a deeper underlying principle that can explain this disparity?

Enter supersymmetry, a theoretical framework that offers a promising solution to the hierarchy problem. At its core, supersymmetry proposes the existence of a new symmetry between fermions (particles with half-integer spin) and bosons (particles with integer spin). This symmetry posits that for every known particle in the Standard Model, there exists a yet-to-be-discovered supersymmetric partner particle with different spin properties.

The introduction of these supersymmetric partners could potentially stabilize the mass of the Higgs boson, a crucial particle responsible for the generation of mass in the universe. Without supersymmetry, the mass of the Higgs boson would receive enormous quantum corrections, leading to a value far beyond what is observed in nature. However, with the inclusion of supersymmetry, these corrections cancel out, resulting in a more reasonable and stable mass for the Higgs boson.

Furthermore, supersymmetry provides a natural candidate for dark matter, a mysterious substance that accounts for a significant portion of the universe's mass. The lightest supersymmetric particle, often referred to as the neutralino, is a leading contender for dark matter. If experimentally proven, this would not only solve the hierarchy problem but also shed light on the elusive nature of dark matter, opening up exciting possibilities for further exploration.

While supersymmetry has yet to be directly observed, its potential implications have captivated the minds of physicists worldwide. Numerous experiments, such as the Large Hadron Collider (LHC), have been conducted to search for signs of supersymmetry. Although no definitive evidence has been found thus far, the quest for supersymmetry continues, driven by the tantalizing prospect of resolving the hierarchy problem and unraveling the mysteries of our universe.

In conclusion, supersymmetry stands as a compelling solution to the hierarchy problem within particle physics. Its introduction not only stabilizes the mass of the Higgs boson but also offers a natural candidate for dark matter. While the search for concrete evidence of supersymmetry is ongoing, the potential implications of this theory have captured the attention of scientists and enthusiasts alike. As we delve deeper into the realm of high-energy particle physics, the pursuit of supersymmetry remains a captivating and essential avenue for understanding the fundamental forces that shape our universe.

Extra Dimensions: Unveiling the Hidden Universe

In the vast and mysterious realm of particle physics, there lies a captivating concept that challenges our very understanding of the universe. This concept, known as extra dimensions, unveils a hidden universe beyond what our senses can perceive. In this subchapter, we will embark on a journey to explore this mind-bending phenomenon and its implications for our understanding of the cosmos.

Imagine a universe where more than the three dimensions we are familiar with—the length, width, and height—exist. According to certain theoretical frameworks, such as string theory and Kaluza-Klein theory, there could be additional dimensions tightly curled up or hidden from our everyday experience. These extra dimensions, existing at a microscopic scale, hold the key to unlocking profound mysteries and revolutionizing our understanding of reality.

But how can we detect these hidden dimensions? The answer lies within the intricate experiments conducted at particle accelerators like the Large Hadron Collider. By colliding particles at extraordinary energies, scientists can probe the fundamental building blocks of matter and potentially reveal the presence of extra dimensions. While direct evidence remains elusive, scientists have devised ingenious experiments and theoretical models that offer tantalizing hints of their existence.

One prominent theory that incorporates extra dimensions is called the "brane-world scenario." According to this theory, our universe resides on a four-dimensional "brane" embedded within a higher-dimensional space. The gravitational force, which appears weaker compared to other fundamental forces, may be diluted in our four-dimensional realm due to its interaction with the extra dimensions. Exploring this theory opens up exciting possibilities for understanding the nature of gravity and why it appears weaker than other forces.

Furthermore, the existence of extra dimensions could shed light on the nature of dark matter and dark energy, two enigmatic components that dominate the universe's composition. These hidden dimensions may be the playground where dark matter particles interact, offering a new perspective on their properties and how they influence cosmic structures.

As we delve deeper into the concept of extra dimensions, we must acknowledge the speculative nature of this field. However, the pursuit of knowledge and understanding is what drives scientific progress. Exploring the possibility of hidden dimensions challenges our preconceptions and expands the boundaries of human knowledge.

In conclusion, the concept of extra dimensions opens up a world of possibilities and challenges our understanding of the universe. Through cutting-edge experiments and theoretical frameworks, scientists are on a quest to unveil this hidden universe. The implications of discovering extra dimensions could revolutionize our understanding of gravity, dark matter, and the fundamental nature of reality itself. So, let us embark on this intellectual adventure, for it is through curiosity and exploration that we push the boundaries of human knowledge and unlock the secrets of the universe.

Chapter 5: Beyond the Standard Model Theories

Grand Unified Theories (GUTs)

In the pursuit of unraveling the mysteries of the universe, scientists have long sought a theory that can explain the fundamental forces of nature. Among the most fascinating and promising contenders in the realm of particle physics are the Grand Unified Theories, or GUTs for short. These theories aim to unify the three fundamental forces – electromagnetism, the weak nuclear force, and the strong nuclear force – into a single, elegant framework.

Addressed to a curious audience of every one, "Beyond the Standard Model: High-Energy Particle Physics for Every Curious Mind" presents an exploration of GUTs, delving into their significance, history, and potential implications for our understanding of the cosmos.

At the heart of GUTs lies the quest for unification. The Standard Model of particle physics, which successfully describes the electromagnetic and weak nuclear forces, has been instrumental in shaping our understanding of the microscopic world. However, it remains incomplete without a unifying theory that encompasses all known forces, including the strong nuclear force. GUTs offer a tantalizing possibility of achieving this unification, providing a more comprehensive framework for understanding the fundamental interactions governing the universe.

To grasp the significance of GUTs, we must delve into the history of their development. Pioneering physicists like Georgi, Glashow, and

Weinberg proposed the first GUT models in the 1970s, building upon the groundbreaking work of theorists preceding them. These early models laid the foundation for further exploration, inspiring a rich tapestry of research and development in the field.

One of the remarkable features of GUTs is their ability to predict the existence of new particles and phenomena. For instance, they propose the existence of X and Y bosons, which mediate interactions between quarks and leptons, offering a potential explanation for the matter-antimatter asymmetry in the universe.

Moreover, GUTs provide insights into the grand unification epoch, a hypothetical period shortly after the Big Bang when all forces were unified into a single force. By studying the predictions of GUTs, scientists can gain valuable insights into the early stages of our universe, shedding light on the profound questions of its origin and evolution.

While GUTs remain theoretical constructs, their potential impact on our understanding of the universe is immense. Advances in experimental techniques, such as those at high-energy particle colliders, offer hope of testing the predictions of GUTs directly. The discovery of proton decay, for example, would provide compelling evidence for the validity of GUTs and open up new avenues for understanding the fundamental nature of our reality.

In "Beyond the Standard Model: High-Energy Particle Physics for Every Curious Mind," we invite you on a journey of discovery, exploring the intricacies of GUTs and their implications for our understanding of particle physics. Whether you are a novice or an enthusiast, this subchapter will equip you with the knowledge and insights to appreciate the significance of GUTs in unraveling the secrets of the universe. Join us as we delve into the frontiers of particle physics, where the quest for unification and understanding takes us beyond the boundaries of the known.

Technicolor: A New Approach to Mass Generation

In the fascinating world of particle physics, scientists constantly strive to push the boundaries of our understanding of the universe. One of the most intriguing areas of research is the search for new models that go beyond the standard model, which has been remarkably successful in explaining the fundamental particles and forces that make up our universe. Among these groundbreaking models is Technicolor, a novel approach to mass generation that has captured the attention of physicists worldwide.

Mass is a fundamental property of particles and plays a crucial role in shaping the universe as we know it. According to the standard model, the Higgs boson is responsible for giving particles mass through the Higgs mechanism. However, Technicolor proposes an alternative mechanism that does not require the existence of a Higgs boson. Instead, it suggests that mass arises from the strong nuclear force, which binds quarks together inside protons and neutrons.

At its core, Technicolor builds upon the concept of quantum chromodynamics (QCD), the theory that describes the strong nuclear force. In QCD, quarks are confined within particles called hadrons due to the strong force. Technicolor extends this confinement idea to the electroweak force, which is responsible for the interactions between particles involved in electromagnetism and the weak nuclear force.

In the Technicolor model, new particles called techniquarks interact through a new strong force called technicolor. These techniquarks form bound states, similarly to how quarks form protons and neutrons. The electroweak force then acts on these bound states, generating mass for the particles in a technicolor condensate. This unique approach provides an elegant solution to the mass generation problem in particle physics.

Although Technicolor offers exciting possibilities for advancing our understanding of mass generation, it is still an active area of research with many questions yet to be answered. Experimental efforts at particle colliders, such as the Large Hadron Collider (LHC), are crucial to test the predictions of Technicolor and search for its signatures. If successful, this groundbreaking model could revolutionize our understanding of the fundamental forces and particles that shape the universe.

Technicolor represents the innovative and creative spirit of particle physics, where scientists continuously push the boundaries of knowledge. By exploring alternative models like Technicolor, we gain deeper insights into the mysteries of the universe and pave the way for future discoveries. As researchers continue to unravel the secrets of mass generation, the world of particle physics promises to captivate and inspire the curious minds of both scientists and enthusiasts alike.

String Theory: Towards a Theory of Everything

In the vast realm of particle physics, scientists have been tirelessly searching for a comprehensive framework that can explain the fundamental forces and particles of nature. One theory that has gained considerable attention and holds the potential to unite all known laws of physics is String Theory. In this subchapter, we delve into the fascinating world of String Theory and its implications in our quest for a Theory of Everything.

String Theory proposes that at the most fundamental level, all particles in the universe are not point-like entities but tiny, vibrating strings. These strings, when vibrating at different frequencies, give rise to the various particles we observe in nature. With this revolutionary concept, String Theory aims to reconcile quantum mechanics and general relativity, two pillars of modern physics that have thus far resisted unification.

One of the most captivating aspects of String Theory is its ability to incorporate gravity into the framework of quantum mechanics. Unlike other theories, String Theory suggests that gravity is not a separate force but emerges from the interactions of these vibrating strings. This tantalizing feature has the potential to resolve the longstanding conflict between our understanding of gravity in the macroscopic world and quantum mechanics at the microscopic level.

Furthermore, String Theory postulates the existence of extra dimensions beyond the familiar four - three spatial dimensions and one dimension of time. These extra dimensions, though hidden from our current observations, play a crucial role in shaping the behavior of the strings. They offer a possible explanation for why gravity appears so weak compared to the other fundamental forces, and they may hold the key to unifying all forces of nature.

However, it is important to note that String Theory is still a work in progress, and many aspects of this theory remain highly speculative. Its mathematical complexity has posed significant challenges, and experimental evidence to support its predictions remains elusive. Nevertheless, String Theory has captivated the minds of physicists worldwide, and its potential to transform our understanding of the universe continues to fuel research and exploration.

In conclusion, String Theory represents a promising avenue toward a Theory of Everything, aiming to unify the fundamental forces and particles of nature within a single framework. While its concepts may be daunting, the profound implications of String Theory for our understanding of the universe make it a captivating subject for anyone curious about particle physics. As scientists continue to unravel the mysteries of String Theory, it is an exciting time to be part of the quest for knowledge at the very fabric of reality.

Dark Matter and Dark Energy: Unsolved Mysteries

In the vast expanse of the universe, there are still countless mysteries waiting to be unraveled. Two of the most intriguing puzzles that continue to baffle scientists are dark matter and dark energy. These enigmatic entities, which make up the majority of the universe, remain elusive and mysterious, challenging our understanding of the cosmos. In this subchapter, we delve into the captivating realm of dark matter and dark energy, shedding light on the ongoing quest to comprehend their nature and significance.

The concept of dark matter emerged as scientists observed that the visible matter we see and interact with makes up only a fraction of the universe's total mass. The rest, approximately 85%, is composed of an invisible and unidentified substance called dark matter. Despite its invisible nature, dark matter exerts gravitational influence on visible matter, holding galaxies together and shaping the large-scale structure of the cosmos. But what is dark matter made of? This question has puzzled physicists for decades, leading to numerous theories and experimental searches. From exotic particles to modifications of gravity, scientists are exploring various avenues to uncover the true identity of dark matter.

While dark matter seems to bind galaxies together, dark energy, on the other hand, is responsible for the accelerated expansion of the universe. First discovered in the late 1990s, dark energy remains an enigma. Unlike anything we have encountered before, dark energy is believed to be a force that permeates all of space, pushing galaxies apart at an ever-increasing rate. Understanding the nature of dark energy is crucial in unraveling the ultimate fate of the universe. Is it a form of energy inherent to the fabric of space itself, or does it arise from some unknown field? These questions continue to challenge scientists, compelling them to explore new theories and devise innovative experiments.

Advancements in particle physics, astrophysics, and cosmology have provided valuable insights into the mysteries of dark matter and dark energy. From the detection of weakly interacting massive particles (WIMPs) to the investigation of cosmic microwave background radiation, scientists are leaving no stone unturned in their quest for answers. The Large Hadron Collider (LHC), the world's most powerful particle accelerator, is also playing a crucial role in the search for dark matter particles.

As we journey deeper into the subatomic world and the vast expanses of the cosmos, the mysteries of dark matter and dark energy continue to captivate our minds. The quest to understand these unsolved mysteries pushes the boundaries of human knowledge and fuels breakthroughs in particle physics. Perhaps one day, we will unlock the secrets of dark matter and dark energy, revealing the hidden truths of our universe and forever transforming our understanding of the cosmos.

Chapter 6: Collider Experiments and Discoveries

Large Hadron Collider (LHC): The World's Most Powerful Accelerator

In the realm of particle physics, the Large Hadron Collider (LHC) stands tall as the world's most powerful accelerator. This subchapter explores the fascinating world of the LHC, its purpose, and its significance in unraveling the mysteries of the universe. Whether you are a curious mind or a particle physics enthusiast, get ready to embark on a journey through the heart of this groundbreaking scientific marvel.

The LHC, situated beneath the Franco-Swiss border near Geneva, Switzerland, is a colossal machine that spans a circumference of 27 kilometers. It is a testament to human ingenuity, engineering prowess, and scientific innovation. Operating at unprecedented energy levels, the LHC allows scientists to collide particles at speeds close to the speed of light. These high-energy collisions simulate conditions similar to those that existed shortly after the Big Bang, providing insights into the fundamental building blocks of the universe.

One of the primary objectives of the LHC is to search for the elusive Higgs boson, a particle that imparts mass to all other particles. In 2012, the discovery of the Higgs boson at the LHC marked a major milestone in particle physics and garnered worldwide attention. This groundbreaking achievement not only confirmed the existence of the Higgs boson but also validated the Standard Model of particle physics.

However, the LHC's quest does not end with the Higgs boson. It also aims to shed light on mysterious phenomena such as dark matter, supersymmetry, and extra dimensions. By colliding particles at unimaginable energies, scientists hope to uncover new particles and forces that lie beyond the Standard Model, providing answers to some of the most perplexing questions about the nature of the universe.

The impact of the LHC extends far beyond the realm of particle physics. Technologies developed for the LHC have found applications in various fields, from medicine to computing. The collaborative nature of the LHC, involving thousands of scientists from around the world, exemplifies the power of international scientific cooperation and highlights the importance of sharing knowledge for the betterment of humanity.

Whether you are a student, a science enthusiast, or simply someone with a curious mind, the LHC offers a gateway to the wonders of particle physics. Delve into the world of high-energy collisions, subatomic particles, and the mysteries of the universe as we explore the fascinating realm of the LHC in this subchapter. Prepare to be amazed, inspired, and captivated by the incredible scientific endeavors taking place at the world's most powerful accelerator.

The Discovery of the Higgs Boson

In the vast field of particle physics, one of the most groundbreaking and anticipated discoveries in recent history was the detection of the elusive Higgs boson. This subchapter delves into the fascinating journey that led to this momentous finding, shedding light on the intricacies of high-energy particle physics.

The Higgs boson, often referred to as the "God particle," is a subatomic particle that plays a crucial role in our understanding of the universe. Its existence was postulated by physicist Peter Higgs and others in the 1960s to explain the origin of mass in fundamental particles. According to the prevailing theory, particles acquire mass by interacting with an all-pervading field called the Higgs field, which is mediated by the Higgs boson.

For decades, scientists tirelessly pursued experimental evidence to confirm the existence of this elusive particle. The quest culminated in the construction of the world's largest and most powerful particle accelerator, the Large Hadron Collider (LHC) at CERN, the European Organization for Nuclear Research. The LHC, a marvel of engineering and scientific ingenuity, allowed researchers to recreate the conditions that existed moments after the Big Bang, facilitating the discovery of the Higgs boson.

The subchapter takes readers on a captivating journey through the intricate experiments conducted at the LHC. It explains the complex concepts such as particle collisions, detectors, and data analysis techniques used to identify the Higgs boson amidst an avalanche of subatomic particles. The chapter elucidates the significance of statistical analysis and the role of countless physicists who worked collaboratively to unveil this groundbreaking discovery.

Furthermore, the subchapter addresses the implications of the Higgs boson discovery in the broader context of particle physics. It explores how this finding has strengthened the Standard Model, the prevailing theory that describes the fundamental particles and their interactions. It also discusses the potential extensions and modifications to the Standard Model, such as supersymmetry and string theory, that the discovery of the Higgs boson has inspired.

Ultimately, this subchapter on the discovery of the Higgs boson captivates readers by unraveling the mysteries of the universe at the smallest scales. It sparks curiosity, fosters understanding, and demonstrates the immense power of human ingenuity and scientific exploration. Whether you are a physicist, a science enthusiast, or simply someone with a curious mind, this subchapter is sure to leave you in awe of the remarkable achievements in the field of particle physics.

Search for Supersymmetry and Other New Physics

In the quest to understand the fundamental building blocks of the universe, particle physicists have gone beyond the Standard Model to explore new frontiers of high-energy physics. This subchapter delves into the fascinating world of the search for supersymmetry and other new physics, shedding light on the cutting-edge research and the implications it holds for our understanding of the cosmos.

Supersymmetry, or SUSY for short, is a theoretical framework that suggests the existence of a whole new set of particles beyond those predicted by the Standard Model. It proposes a symmetry between particles that have integer spins (like protons and electrons) and those with half-integer spins (like quarks and leptons). This symmetry could provide answers to many unanswered questions, such as the nature of dark matter and the unification of fundamental forces.

The Large Hadron Collider (LHC), the world's most powerful particle accelerator, has been instrumental in the search for supersymmetry. Physicists have been eagerly analyzing the data produced by the LHC experiments, hoping to catch a glimpse of these elusive supersymmetric particles. The discovery of supersymmetry would revolutionize our understanding of the universe, but so far, no direct evidence has been found.

However, the absence of evidence does not necessarily mean the absence of supersymmetry. Researchers are continuously refining their theoretical models and experimental techniques to probe deeper into the unexplored territories of particle physics. They are exploring alternative scenarios, such as hidden sectors and non-standard SUSY models, to expand the possibilities and increase the chances of discovering new physics.

The search for supersymmetry is not limited to the LHC alone. Other experiments, such as the search for dark matter particles and neutrino oscillations, also contribute to our understanding of new physics beyond the Standard Model. These experiments provide valuable insights into the mysteries of the universe and push the boundaries of our knowledge.

While the search for supersymmetry remains ongoing, it represents just one facet of the broader quest to unravel the mysteries of the universe. Particle physicists are driven by an insatiable curiosity to explore the unknown and understand the fundamental laws that govern our existence. Their research not only impacts our understanding of particle physics but also has profound implications for cosmology, astrophysics, and even the technological advancements of tomorrow.

In conclusion, the search for supersymmetry and other new physics is an exciting and vibrant field of study in particle physics. It represents our relentless pursuit to uncover the secrets of the universe and expand the frontiers of human knowledge. As we continue to delve deeper into the mysteries of the cosmos, we inch closer to a more comprehensive understanding of the fundamental nature of reality.

Future Colliders: Pushing the Energy Frontier

In the ever-evolving field of particle physics, scientists are constantly seeking to unravel the mysteries of the universe. The quest to understand the fundamental building blocks of matter and the forces that govern their interactions has led to groundbreaking discoveries over the years. However, there is still much to learn, and the next frontier in this exciting journey lies in the development of future colliders.

Colliders are powerful machines that accelerate particles to nearly the speed of light and smash them together, creating conditions similar to those that existed shortly after the Big Bang. By studying the debris of these collisions, scientists can unlock the secrets of the universe and gain insights into the nature of matter itself.

The current crown jewel of particle colliders is the Large Hadron Collider (LHC) at CERN, which made headlines in 2012 with the discovery of the Higgs boson. This groundbreaking finding validated the existence of the Higgs field, which gives particles mass and is a cornerstone of the Standard Model of particle physics. However, the LHC is just the beginning.

Future colliders are being proposed to push the energy frontier even further, enabling scientists to explore new realms of physics and potentially uncover phenomena that go beyond the Standard Model. One such proposed project is the International Linear Collider (ILC), which aims to collide electrons and positrons. Unlike the LHC, which accelerates protons, the ILC would provide a cleaner collision environment, allowing for precise measurements and potential discoveries.

Another ambitious project on the horizon is the Compact Linear Collider (CLIC), which envisions accelerating particles to even higher energies than the ILC. This would enable scientists to study the properties of particles with unprecedented precision and potentially shed light on the mysteries of dark matter and dark energy.

These future colliders are not just the realm of scientists; they have implications for everyone. The discoveries made at these machines have the potential to revolutionize technology, medicine, and our understanding of the universe. From advancements in particle detectors to breakthroughs in data analysis and computing, the technological spin-offs from these projects can benefit society as a whole.

The study of particle physics is not limited to a niche audience. It captivates the curious minds of people from all walks of life who are fascinated by the mysteries of the universe. Whether you are a student, a scientist, or simply someone with a thirst for knowledge, the future colliders represent the next chapter in our collective journey towards understanding the cosmos.

So, join us as we embark on this thrilling adventure, pushing the energy frontier and unraveling the secrets of the universe. Together, we can explore the unknown and satisfy our curiosity about the fundamental nature of our existence. The future colliders await, and they have the potential to change our understanding of the world forever.

Chapter 7: Applications of High-Energy Particle Physics

Medical Applications: Cancer Treatment and Imaging

In the fascinating world of particle physics, where scientists delve into the mysteries of the universe at the tiniest scales, there lies a realm of knowledge that holds great potential for medical applications. Beyond the Standard Model, the realm of high-energy particle physics, offers groundbreaking insights and technologies that can revolutionize cancer treatment and imaging. In this subchapter, we will explore how the discoveries made in particle physics can be harnessed to benefit every individual, shedding light on the niches of particle physics and addressing a broad audience.

Cancer, one of the deadliest and most prevalent diseases in the world, has long been a challenge for medical science. However, particle physics has paved the way for innovative approaches to cancer treatment. One such technique is proton therapy, a radiation therapy that uses accelerated protons to target cancer cells precisely while minimizing damage to healthy tissues. Protons, being charged particles, can be controlled to deposit their energy precisely at the tumor site, with minimal scatter. This leads to reduced side effects and enhanced efficiency in eradicating cancer cells, offering new hope to patients battling this formidable disease.

Moreover, the knowledge gained from particle physics has also revolutionized the field of medical imaging. Positron emission tomography (PET) is a prime example of this. By harnessing the principles of particle physics, PET scans can detect radioactive isotopes injected into the body and produce detailed images of metabolic activity. This non-invasive technique enables early detection of cancer, aiding in accurate diagnosis and effective treatment planning. Furthermore, the combination of PET with other imaging modalities like magnetic resonance imaging (MRI) and computed tomography (CT) provides a comprehensive understanding of the disease, facilitating personalized medicine and improving patient outcomes.

The advancements in particle physics have also contributed to the development of targeted therapies, such as immunotherapy and gene therapy. By understanding the fundamental particles and forces governing nature, scientists have unraveled the intricate mechanisms of the human immune system and genetic makeup. This knowledge has led to the development of groundbreaking treatments that harness the body's own defenses or modify genetic material to combat cancer cells specifically, minimizing harm to healthy cells.

In conclusion, the world of particle physics holds immense potential for the field of medicine, particularly in cancer treatment and imaging. Proton therapy, PET scans, targeted therapies, and many other innovations are transforming the way we fight against cancer. By bridging the gap between particle physics and medical science, we can bring cutting-edge technologies and insights to the forefront of healthcare, benefiting everyone. As we continue our journey beyond the Standard Model, the impact of particle physics in our lives becomes increasingly evident, offering hope for a healthier future for all.

Particle Physics and Cosmology: Understanding the Universe's Evolution

The subchapter titled "Particle Physics and Cosmology: Understanding the Universe's Evolution" delves into the fascinating connection between the two fields of study and how they contribute to our understanding of the universe's evolution. This subchapter aims to engage a wide audience, from those with a cursory interest in science to enthusiasts of particle physics.

At the heart of our exploration lies the question: How did the universe come to be? To answer this fundamental query, we must embark on a journey into the realms of particle physics and cosmology. Particle physics investigates the fundamental building blocks of matter and the forces that govern them, while cosmology explores the structure and evolution of the universe as a whole.

In this subchapter, we will explore the intertwined nature of these disciplines, highlighting their shared concepts, methodologies, and discoveries. One of the central themes we will touch upon is the Big Bang theory, which provides a framework for understanding the universe's birth and subsequent expansion. We will discuss how particle physics plays a crucial role in unraveling the mysteries of the early universe, taking us back to moments just fractions of a second after the Big Bang.

As we delve deeper, we will explore the concept of dark matter and dark energy, two enigmatic entities that dominate the composition and dynamics of the universe. Particle physicists are actively searching for evidence of these elusive particles, which are thought to make up a significant portion of the cosmos. We will discuss the ongoing experiments and theoretical models that aim to shed light on these cosmic enigmas.

Furthermore, we will unravel the intricate connection between particle physics and the formation of galaxies, stars, and other celestial structures. By studying the properties of subatomic particles, scientists can simulate the conditions of the early universe and gain insights into the mechanisms that led to the formation of the cosmic structures we observe today.

Throughout this subchapter, we will present the latest advancements in both particle physics and cosmology, discussing the cutting-edge experiments and theoretical frameworks that continue to shape our understanding of the universe. From the Large Hadron Collider to cosmic microwave background radiation studies, we will explore the vast array of tools and techniques employed by scientists in these fields.

By the end of this subchapter, readers will gain a deeper appreciation for the intricate relationship between particle physics and cosmology. They will understand how the exploration of subatomic particles contributes to our understanding of the universe's evolution and the fundamental laws that govern its existence. Whether you are a novice or an avid follower of particle physics, this subchapter promises to captivate your curiosity and expand your knowledge of the cosmos.

Technology Spin-offs: From Particle Detectors to Everyday Life

In the vast realm of particle physics, where scientists delve into the mysteries of the universe at its most fundamental levels, there exists a fascinating intersection between cutting-edge technology and the everyday world we inhabit. Particle detectors, the ingenious devices designed to capture and measure the elusive particles that make up our universe, have not only revolutionized our understanding of physics but have also spawned a plethora of unexpected technological spin-offs that have found their way into our daily lives.

One of the most remarkable examples of this technology transfer is the development of medical imaging devices. The advanced imaging techniques used in particle detectors, such as positron emission tomography (PET) and magnetic resonance imaging (MRI), have been adapted and refined to create powerful tools for diagnosing diseases and monitoring treatments. These technologies enable doctors to visualize the internal structures of our bodies with unprecedented precision, assisting in the early detection of ailments and guiding interventions. Thanks to the breakthroughs in particle physics, medical imaging has become an invaluable asset in modern healthcare, saving countless lives and improving patient outcomes.

Furthermore, the quest for understanding the fundamental building blocks of matter has propelled advancements in materials science. Particle accelerators, used to accelerate particles to near-light speeds for collision experiments, require cutting-edge materials capable of withstanding extreme conditions. The development of resilient, high-performance materials to build these accelerators has found applications in various industries, ranging from aerospace to transportation. The lightweight yet robust materials used in particle accelerators have significantly contributed to the design of fuel-efficient vehicles, stronger and safer aircraft structures, and even more durable sports equipment.

Beyond the realms of medicine and materials, particle physics technology has also infiltrated the realm of everyday consumer electronics. The invention of the silicon microstrip detector, initially designed to track particles in high-energy experiments, has paved the way for the miniaturization of electronic devices. This revolutionary technology has revolutionized the smartphone industry, allowing for the creation of smaller, more powerful, and energy-efficient devices that fit comfortably in our pockets. Moreover, the development of high-speed data transfer techniques, a necessity in particle physics experiments, has directly influenced the evolution of internet connectivity, enabling faster and more reliable communication across the globe.

In conclusion, the technological advancements spurred by particle physics have permeated every aspect of our lives, from healthcare to materials science and consumer electronics. The ingenious devices and techniques developed for unraveling the mysteries of the universe have had far-reaching implications, benefiting humanity in ways unimaginable. As we continue to explore the frontiers of particle physics, we can only anticipate further exciting spin-offs that will continue to shape and transform our everyday lives.

Chapter 8: Open Questions and Future Directions

Neutrino Oscillations: Shedding Light on Neutrino Properties

In the vast realm of particle physics, one enigmatic entity has captured the curiosity of scientists and enthusiasts alike - the neutrino. These elusive particles have long remained mysterious due to their elusive nature and their ability to interact weakly with matter. However, a groundbreaking discovery in the late 20th century, known as neutrino oscillations, has revolutionized our understanding of these subatomic particles and shed new light on their properties.

Neutrino oscillations refer to the phenomenon where neutrinos change their flavor as they travel through space. Initially, three types of neutrinos were identified - electron, muon, and tau neutrinos. However, experiments conducted over several decades have revealed that these neutrinos can morph into each other, indicating the presence of neutrino oscillations. This remarkable discovery has not only confirmed the existence of neutrino masses, but it has also challenged the traditional Standard Model of particle physics.

The book "Beyond the Standard Model: High-Energy Particle Physics for Every Curious Mind" delves into the captivating world of neutrino oscillations. Written for everyone, from passionate enthusiasts to aspiring physicists, this subchapter aims to unravel the complexities surrounding neutrino properties and their oscillations.

The subchapter begins by providing a comprehensive overview of the Standard Model and its limitations in explaining neutrino oscillations. It then delves into the theoretical framework behind neutrino oscillations, exploring concepts such as neutrino mixing, flavor eigenstates, and mass eigenstates. The book elucidates the mathematical formalism required to understand neutrino oscillations, ensuring accessibility for readers from all backgrounds.

Furthermore, the subchapter explores the experimental evidence supporting neutrino oscillations, including landmark experiments like Super-Kamiokande and the Sudbury Neutrino Observatory. It highlights the collaborations between scientists worldwide, demonstrating the importance of international cooperation in unraveling the secrets of the neutrino.

Moreover, the book emphasizes the implications of neutrino oscillations beyond particle physics. From astrophysics to cosmology, neutrino oscillations have far-reaching consequences, influencing our understanding of the early universe, supernovae, and even the existence of dark matter.

In conclusion, this subchapter on neutrino oscillations in "Beyond the Standard Model: High-Energy Particle Physics for Every Curious Mind" offers an engaging exploration of these elusive particles and their oscillatory behavior. By presenting the theoretical and experimental foundations of neutrino oscillations, this subchapter provides readers with a comprehensive understanding of this fascinating field. Whether you are a seasoned physicist or simply curious about the mysteries of the universe, this subchapter is sure to captivate your mind and deepen your appreciation for the wonders of particle physics.

Baryogenesis: Explaining the Matter-Antimatter Asymmetry

In the captivating world of particle physics, one of the most intriguing mysteries lies in the asymmetry between matter and antimatter. Why does our universe seem to be predominantly composed of matter, while antimatter remains scarce? This enigmatic phenomenon is known as the matter-antimatter asymmetry, and its explanation lies in the realm of baryogenesis.

Baryogenesis, derived from the Greek words "barys" meaning heavy and "genesis" meaning creation, refers to the process that gave rise to the excess of matter over antimatter in our universe. To understand this concept, we must delve into the early stages of our universe's formation, specifically during the epoch known as the Big Bang.

During the Big Bang, the universe experienced a rapid expansion and intense energy fluctuations. As a result, particles and antiparticles were constantly being created and annihilated. However, something peculiar occurred during this tumultuous period – a tiny asymmetry emerged, favoring the creation of matter particles over antimatter particles.

To comprehend this asymmetry, scientists have proposed various theoretical frameworks, such as the Sakharov conditions. These conditions stipulate three requirements for baryogenesis to occur: violation of baryon number conservation, violation of charge-parity (CP) symmetry, and departure from thermal equilibrium.

The violation of baryon number conservation suggests that the total number of baryons (protons and neutrons) in the universe is not a constant. This allows for the possibility of creating more baryons than antibaryons, leading to an imbalance in matter and antimatter.

The second condition, violation of CP symmetry, implies that there is a fundamental difference in the behavior of matter and antimatter particles. This phenomenon has been observed in certain particle interactions, providing a plausible explanation for the matter-antimatter asymmetry.

Finally, the departure from thermal equilibrium is crucial for baryogenesis. During the early universe, when particle interactions were frequent and energetic, the universe was in thermal equilibrium. However, as the universe expanded and cooled down, it underwent a phase transition, causing a departure from thermal equilibrium. This departure allowed for the creation of asymmetry between matter and antimatter.

Understanding baryogenesis is not only crucial for unraveling the mysteries of our universe's origins but also for shedding light on fundamental questions in particle physics. Exploring the intricacies of baryogenesis allows us to grasp the delicate balance between matter and antimatter, and how this balance shaped the world as we know it today.

In conclusion, baryogenesis serves as a fascinating subfield within particle physics, explaining the matter-antimatter asymmetry that pervades our universe. By violating baryon number conservation, CP symmetry, and departing from thermal equilibrium, scientists have begun to unravel the mystery behind why our universe is predominantly composed of matter. Through continued research and experimentation, we hope to gain deeper insights into the origins of our universe and the fundamental nature of the particles that make up our reality.

Neutrinoless Double Beta Decay: Probing the Nature of Neutrinos

The quest to understand the fundamental nature of the universe has led scientists to explore the mysterious realm of particle physics. At the heart of this exploration lies the study of neutrinos - elusive particles that hold the key to unraveling the mysteries of the cosmos. In this subchapter, we delve into the intriguing phenomenon of neutrinoless double beta decay, a process that has the potential to revolutionize our understanding of particle physics.

Neutrinos are fascinating particles that are produced in abundance by nuclear reactions, such as those occurring within the Sun or during radioactive decay. They are incredibly lightweight and have the peculiar ability to change from one flavor to another as they travel through space. However, their ghostly nature and weak interactions make them extremely challenging to detect and study.

One of the most important questions surrounding neutrinos is whether they are their own antiparticles, a concept known as Majorana particles. Neutrinoless double beta decay provides a unique window into this mystery. This rare process, which involves the simultaneous decay of two neutrons within an atomic nucleus, could only occur if neutrinos are Majorana particles.

The detection of neutrinoless double beta decay would have profound implications for our understanding of the universe. It would confirm the violation of lepton number conservation, a fundamental symmetry in particle physics. This violation would shed light on why there is an asymmetry between matter and antimatter in the universe, offering a potential explanation for our existence.

The search for neutrinoless double beta decay is an ongoing endeavor, with numerous experiments being conducted worldwide. These experiments employ cutting-edge technologies, such as ultra-pure detectors and shielding to minimize background noise. By meticulously analyzing the data collected from these experiments, scientists hope to catch a glimpse of this elusive phenomenon.

The discovery of neutrinoless double beta decay would not only confirm the nature of neutrinos but also open doors to new physics beyond the Standard Model. It would provide valuable insights into the origin of mass and the hierarchy problem, as well as offer a deeper understanding of the fundamental forces that govern our universe.

In conclusion, neutrinoless double beta decay is a captivating process that holds immense potential for advancing our knowledge of particle physics. Its detection would not only confirm the existence of Majorana neutrinos but also revolutionize our understanding of the universe at its most fundamental level. The ongoing experimental efforts in this field are a testament to human curiosity and the relentless pursuit of scientific knowledge.

The Future of High-Energy Particle Physics: Challenges and Opportunities

In the ever-evolving field of high-energy particle physics, the thirst for knowledge and the quest to unravel the mysteries of the universe continue to push the boundaries of human understanding. Particle physics, often referred to as the study of the fundamental building blocks of matter and their interactions, has come a long way since the inception of the Standard Model. However, as we delve deeper into the complexities of the universe, new challenges and exciting opportunities lay ahead.

One of the greatest challenges facing high-energy particle physics is the need for ever more powerful and sophisticated experimental facilities. Particle accelerators, such as the Large Hadron Collider (LHC), have been instrumental in probing the fundamental particles and forces that govern our universe. However, to explore new frontiers, we require even larger and more advanced machines. The construction and operation of these next-generation accelerators pose significant technical and financial challenges, but they also hold immense potential for groundbreaking discoveries.

Another challenge lies in the need for international collaboration. Particle physics is a global endeavor, with scientists from different countries working together to conduct experiments and analyze data. The future of high-energy particle physics relies on fostering strong partnerships and promoting open exchange of ideas and resources. Collaboration not only helps in sharing the enormous costs associated with cutting-edge research but also brings together diverse perspectives and expertise, leading to innovative breakthroughs.

Advances in computing and data analysis also present both a challenge and an opportunity. Modern particle physics experiments generate an enormous amount of data, and analyzing this vast sea of information requires sophisticated algorithms and computational power. The development of faster, more efficient data processing techniques is essential for making sense of the intricate patterns and phenomena observed in particle collisions.

While challenges lie ahead, high-energy particle physics also presents exciting opportunities. The search for new particles and forces beyond the Standard Model, such as dark matter and supersymmetry, is a thrilling frontier. These discoveries could revolutionize our understanding of the universe and have profound implications for fields ranging from cosmology to technology.

Moreover, particle physics has a tremendous impact on society beyond its scientific contributions. The technological advancements driven by particle physics research have led to significant breakthroughs in medical imaging, materials science, and energy production. By investing in high-energy particle physics, we are investing in our future, fostering innovation, and driving economic growth.

The future of high-energy particle physics holds immense promise, but it is not without its challenges. By embracing these challenges, fostering collaboration, and pushing the boundaries of our knowledge, we can continue to unlock the secrets of the universe and pave the way for a better tomorrow.

Chapter 9: Conclusion

Recapitulation of Key Concepts

In this subchapter, we will take a moment to recapitulate the key concepts discussed throughout the book, "Beyond the Standard Model: High-Energy Particle Physics for Every Curious Mind." With the aim of addressing a diverse audience, ranging from seasoned physicists to individuals with a general interest in particle physics, this recap will serve as a concise reminder of the fundamental concepts explored in this book.

First and foremost, we have embarked on a journey to explore the realm of particle physics, a field that delves into the fundamental building blocks of the universe and the forces that govern them. We have discussed the Standard Model, the prevailing theory that describes the electromagnetic, weak, and strong nuclear forces, as well as the elementary particles that compose matter.

Within the Standard Model, we have examined the fundamental particles, such as quarks, leptons, and gauge bosons, which play a pivotal role in understanding the structure of matter. Furthermore, we have explored the concept of symmetries and how they shape the fundamental interactions between these particles.

Moving beyond the Standard Model, we have ventured into the realm of high-energy particle physics, where we have discussed the existence of particles beyond those predicted by the Standard Model. We have explored the evidence for dark matter, a mysterious substance that constitutes a significant portion of the universe yet eludes direct detection.

Additionally, we have delved into the concept of supersymmetry, a theoretical framework that proposes a symmetry between particles with integer and half-integer spins. This theory provides a potential explanation for the nature of dark matter and could pave the way for a more complete understanding of the fundamental forces.

Throughout this book, we have also emphasized the vital role of particle accelerators and detectors in advancing our knowledge of particle physics. From the Large Hadron Collider (LHC) to cutting-edge experiments, these scientific instruments have allowed us to probe the fundamental structure of matter and search for new particles beyond the Standard Model.

In conclusion, "Beyond the Standard Model: High-Energy Particle Physics for Every Curious Mind" has provided an overview of the key concepts in particle physics. Whether you are an enthusiast with a general interest in the subject or an aspiring physicist, this book aims to ignite curiosity and lay the groundwork for further exploration into the fascinating world of particle physics.

The Impact of High-Energy Particle Physics

Particle physics, a branch of physics that deals with the smallest building blocks of matter and the fundamental forces that govern their interactions, has had a profound impact on our understanding of the universe. High-energy particle physics, in particular, has pushed the boundaries of our knowledge and revolutionized our understanding of the fundamental laws of nature.

One of the major impacts of high-energy particle physics is the discovery of new particles. Through experiments conducted at particle accelerators such as the Large Hadron Collider (LHC), scientists have been able to uncover particles that were previously unknown to us. For example, the discovery of the Higgs boson at the LHC in 2012 confirmed the existence of the Higgs field, which is responsible for giving mass to other particles. This discovery was a significant milestone in our understanding of the fundamental nature of matter and has opened up new avenues of research.

In addition to discovering new particles, high-energy particle physics has also led to the development of new technologies. The construction and operation of particle accelerators require cutting-edge engineering and innovative technologies. These advancements often find applications in other fields, such as medical imaging, materials science, and even computing. For instance, the development of superconducting magnets for particle accelerators has led to breakthroughs in magnetic resonance imaging (MRI) technology, making it a vital tool in modern healthcare.

Furthermore, high-energy particle physics has contributed to our understanding of the early universe. By recreating the extreme conditions that existed shortly after the Big Bang, scientists can study the fundamental forces and particles that governed the universe's evolution. This research provides insights into the origins of the universe, the nature of dark matter and dark energy, and the possible existence of extra dimensions.

Moreover, high-energy particle physics has fostered international collaboration among scientists from different countries and backgrounds. The scale of experiments and the complexity of the data analysis require teams of researchers working together. This collaborative effort promotes the exchange of ideas and expertise, leading to advancements not only in particle physics but also in various other scientific disciplines.

In conclusion, high-energy particle physics has had a far-reaching impact on our understanding of the universe, the development of new technologies, and the promotion of international collaboration. It has allowed us to delve deeper into the fundamental nature of matter and energy, unravel the mysteries of the early universe, and contribute to technological advancements that benefit society as a whole. Whether you are an aspiring physicist or simply curious about the wonders of the universe, high-energy particle physics offers an exciting and rewarding journey of discovery.